CALCULUS LABORATORIES WITH MATHEMATICA

Volume 1

Michael G. Kerckhove and Van C. Nall

University of Richmond

McGRAW-HILL, INC.

New York   St. Louis   San Francisco   Auckland   Bogotá   Caracas
Lisbon   London   Madrid   Mexico   Milan   Montreal   New Delhi
Paris   San Juan   Singapore   Sydney   Tokyo   Toronto

This book requires Mathematica version 2.0.
Mathematica is a registered trademark of Wolfran Research, Inc.

CALCULUS LABORATORIES WITH MATHEMATICA, Volume 1

1 2 3 4 5 6 7 8 9 0   MAL MAL   9 0 9 8 7 6 5 4 3 2

ISBN 0-07-034220-2

The editor was Maggie Lanzillo;
the production supervisor was Denise L. Puryear.
Malloy Lithographing, Inc., was printer and binder.

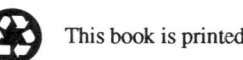

Preface

This collection of computer-based exercises for calculus has been developed over a period of nearly three years and has been used in eight sections of the first semester course during that time. These exercises form a valuable component of our course, and we feel they have been sufficiently tested and refined so that others may benefit from them as well. To assist you in adapting these exercises to your calculus course, this preface contains information about the motivations, aspirations, and practical considerations that have guided their development and about the technical and logistical environment in which they have been used successfully.

Our primary goal is to bring more meaning to the material learned by our students. Many students become proficient at manipulating the symbols of calculus, but have real difficulty in understanding the significance of their calculations or in drawing meaningful conclusions from them. Computers can be used to highlight features of problems other than symbolic manipulation. Studying functions in a laboratory setting where they are asked to write about their observations helps students become more mathematically articulate and precise.

We believe that we have discovered some new and useful ways to use the computer, and we are attempting to expand the use of computers in the teaching of mathematics. We try to avoid inventing problems as an excuse to use the software, or forcing the software to be used to solve problems where it does not enhance learning. Therefore, these computer exercises are not intended to be comprehensive. They were designed to be used with a standard textbook, and in a course which would leave students prepared to take the sequel with students who have had a standard course without the use of computers. They can be used either as a separate laboratory component, or as a replacement for some of the usually assigned problems. Each exercise set can be completed in approximately two hours, except the notebook on max-min problems which, if all the problems are assigned, will take closer to four hours. Our students report that they spend an average of 5 to 6 hours each week on all course assignments.

We have found that students learn to use Mathematica by using it. The documentation we have provided for Mathematica is minimal, but has proven to be sufficient. We express functions and pose questions in as close to standard notation as possible, leaving the task of translation into Mathematica notation to the students. We have found that this approach quickly produces competent and independent users of Mathematica.

You will need to provide relevant information to students about matters particular to your lab setup that we cannot foresee. We do have some suggestions pertaining to that setup.

You must have machines that you are certain will run Mathematica version 2.0 with the Notebooks front end. The number of machines will depend on the patterns of use you decide to establish.

Our students work in pairs. This provides more efficient use of the computers and of the student's time, and we believe that the student's ability to write coherent explanations benefits especially from working with a partner. Groups of three have not worked well. If we had a lab period and enough equipment, we would have students work some sets of exercises in pairs and some individually.

We provide a paper copy of the exercises which is easier to read and provides a backup to the digital version located on their computers. Students type their work into new cells that they open between the cells which contain the questions. Their answers may contain graphics and other Mathematica output, as well as text cells. Students save their work often and, upon completion, the work is transferred to a file owned by the instructor. Only the notebook "Graphing " requires printing, which should otherwise be discouraged.

We would be pleased to receive your comments. Good luck.

Michael G. Kerckhove,
Van C. Nall

Department of Mathematics and Computer Science
University of Richmond
Richmond, Virginia 23173

The authors wish to acknowledge support from the National Science Foundation and Wolfram Research, Inc.

Contents

Each set of exercises, also called a "notebook," was designed to accompany a section of the textbook <u>Calculus and Analytic Geometry</u> , fifth edition, by Stein and Barcellos. The last notebook was designed as a special project for ambitious students; it is much more open-ended and difficult than the others.

Using Mathematica..1

Notebooks   *Section of Text*

A. Plot and Solve...Chap. 2................................ 3

B. Asymptotes of Rational Functions......................2.5..........................5

C. Limits.. 2.7................................8

D. Introduction to Derivatives................................3.1..............................10

E. Tangent Lines...3.2..............................13

F. Newton's Method..4.6................................15

G. Derivatives and Estimates I...............................4.1..............................19

H. Derivatives and Estimates II.............................. 4.10..............................22

I. Graphing...4.5..............................24

J. Max-Min Problems..4.7.............................25

K. Some Sums..5.2.............................31

L. Riemann Sums...5.3..............................35

M. Tower of Powers..6.5.............................39

A Minimal List of Mathematica Commands...40

Using Mathematica

Predefined Functions in Mathematica

Mathematica has many predefined functions, including

Abs[x] -- Absolute Value
Cos[x] -- Cosine
Sin[x] -- Sine
Sqrt[x] -- Square Root
ArcTan[x] -- Inverse Tangent

A more complete list can be found in the Mathematica book. The name of each built-in function begins with an upper case letter and Mathematica uses square brackets in place of the more common round brackets for function arguments.

Defining Your Own Functions

Mathematica allows you to define your own functions. Here is an example of how to define the squaring function.

```
f [x_]  :=  x^2
```

You can use whatever names you like; however, you should avoid using names that start with upper case letters to distinguish functions you define from Mathematica's built-in functions. In this case, we used the rather unimaginative name "f".

You should notice two other things about this definition. We used the ":=" assignment operator instead of "=". Either one would have worked; the difference between the two is subtle. For now, just remember that you will usually want to use ":=".

The other thing you should notice is that on the left side of the definition we wrote the variable as "x_", with an underscore. In effect, the above definition tells Mathematica that **f[** *anything* **]** is anything squared. Without the underscore, that is f[x] := x^2, Mathematica understands that f[x] is x^2 but does not know what to do with **f[2]** or **f[y]**. The expression "**f[x_] := x^2**" defines a rule, whereas the expression "**g[x] := x^2**" merely assigns an expression to g[x].

1

Using Plot

The basic command for plotting graphs of functions is **Plot**. The syntax is
Plot[f[x], {x, *xmin*, *xmax*}] where f is the name of the function, x is the independent
variable, and xmin to xmax is the range over which x is evaluated. Here is an example.

```
Plot[Cos[x], {x, 0, 2 Pi}]
```

You can plot several functions at once by listing their names in braces.

```
Plot[{Sin[x], Sin[2x]}, {x, 0, 2 Pi}]
```

Solving Equations

Mathematica has two principal tools for solving equations. The first is the **Solve**
command with syntax

```
Solve[lhs == rhs , variable],
```

where lhs and rhs are any sort of expression involving the variable. **Solve** will always
work if rhs and lhs are polynomials of degree five or less. Here is an example.

```
Solve[x^4 + x - 1 == x , x]
```

If **Solve** cannot find the solution, you can obtain a numerical approximation with the
command **FindRoot** which has syntax

```
FindRoot[ lhs == rhs , {variable, rough guess}]
```

FindRoot will find an approximate value for a solution close to your rough guess. Here is
an example.

```
FindRoot[Sin[x] + Cos[x] == 0 , {x, 2}]
```

The results of **FindRoot** will be in decimal notation, and the results of **Solve** will be exact
numbers which may be hard to locate. You can use **N[%]** to convert the output of **Solve**
into decimal form.

Plot and Solve

Your computer algebra system includes a command to plot graphs of functions, which has the syntax

$$\texttt{Plot[f[x],\{x,a,b\}]}$$

and a command to solve equations, which has the syntax

$$\texttt{Solve[f[x] == a , x]}$$

The exercises below illustrate some uses for these commands.

Exercises

1. Solve each equation for x. Get an exact solution if possible. Use **N[%]** to get numerical approximations to each solution. For each equation, how many solutions are real?

    a. $3x^2 + 2x - 5 = 0$

    b. $x^3 - 4x^2 + 3x - 5 = 0$

    c. $(2x - 4)^{1/2} - (3x + 4)^{1/2} = -2$

2. Derive the quadratic formula by solving the equation $ax^2 + bx + c = 0$ for x. Try to get Mathematica to put the result in the standard form. (Hint: use **Solve** and **Simplify**.)

3.   Plot the graph of each function below. Be sure to select an interval and plot range that illustrates all important features. You will probably have to experiment some before you are satisfied. Find all x-intercepts using **Solve**, and determine whether the results of solve and the plot of the function seem to agree.

It is good practice to enter **Clear[f,g,h]** at the start of these exercises.

    a. $f(x) = (x^2 - 2)(x^3 - x)$ Find all x-intercepts.

    b. $g(x) = (x^2 - 3x + 2)(x^3 + x + 1) - 1$ Find all x-intercepts. Use **Solve**.

    c. $h(x) = x^5 - 8x^4 + 11x^3 + 56x^2 - 180x + 144$ Find all x-intercepts. Be careful.

The next three exercises deal with transformations of graphs. Experiment using Mathematica to gain an understanding of how a simple change in the formula for a function influences the appearance of the corresponding graph. Write a thorough presentation of your results, combining text and pictures to adequately describe your work and to discuss the guiding principles you have discovered.

It is good practice to enter **Clear[f,g]** at the start of these exercises.

4. If f is a function then the graph of f(x - a), where a is a constant, is called a horizontal translation of the graph of f.

    a.   Let $f(x) = x^2$. Plot the graph of f(x), f(x + 1) and f(x - 1).

    b.   Let $g(x) = x^3 - x$. Plot the graph of g(x), g(x + 1), and g(x - 1).

    c.   Describe, in general, how the graph of a function, f(x - a), changes as a is changed. Be sure to consider both positive and negative values of a.

5. Describe, in general, how the graph of a function, f(a x), changes as a is changed. Be sure to consider both positive and negative values of a. Consider values of a between 0 and 1, as well as values that are greater than 1. This transformation is called a horizontal scale change. Experiment with plots based on the two functions in problem 4.

6. Plot the function f(x) = a sin(bx + c) with different values for a,b, and c, and describe how the graph changes as each of a, b, and c are changed.

Mathematica Commands Needed:

```
Clear[ variables ]

Plot[ f[x], {x, a, b} ]

Simplify [ expression ]

Solve[ f[x] == a , x ]
```

Asymptotes of Rational Functions

In this section you will explore the graphs of rational functions: functions of the type $f(x) = p(x)/q(x)$. The **Apart** command will factor the denominator, $q(x)$, and express $f(x)$ as a sum of a polynomial with degree equal to the degree of $p(x)$ minus the degree of $q(x)$, and a sequence of rational functions whose denominators are the factors of $q(x)$, and whose numerators have degree less than their denominators. Here is an example.

```
f[x_]:= (x^6 - x^2 + 1)/(x^5 - 6x^4 + 12x^3 - 12x^2 +
        11x - 6)
```

```
Apart[f[x]]
```

```
         721                61                1
6 + -------------  - ------------- + ------------- + x -
    20 (-3 + x)      5 (-2 + x)      4 (-1 + x)

         x
    -------------
         2
    10 (1 + x )
```

The graph of $f(x)$ will have vertical asymptotes at $x = 1$, $x = 2$, and $x = 3$, and it will have an oblique asymptote along the line $y = x + 6$. Several plots with different domains and ranges are needed to see all of these features. Try it.

Exercises

1. Use the result of **Apart[f[x]]** to explain why the function $f(x)$ above has an oblique asymptote along $y = x + 6$.

2. For each function listed below enter the definition of $f(x)$, and then enter the command **Apart[f[x]]**. (Make sure you **Clear[f]** before entering each new function definition.) Plot the graph of each function. Experiment with the domain and range in order to get a graph or several graphs which best display the major features of the function including all vertical, horizontal, and oblique asymptotes. Use the result of **Apart[f[x]]** to explain the presence or absence of each type of asymptote.

   a. $f(x) = 1/(x + 1)^2$

5

b. $f(x) = (4x + 3)/(3x - 4)$

c. $f(x) = (x^2 - 7x + 5)/(2x^3 - 5x^2 - 8x + 6)$

d. $f(x) = x^3/((x^2+1)(x^2 + 3))$

e. $f(x) = (x^4 + x + 1)/(x^2 - 1)$

3. Enter the definition of the mysterious compiled function g.

```
Clear[g]
```

```
g = CompiledFunction[{_Real}, {0, 2, 7, 0},
      {{1, 17}, {4, 1, 0}, {12, 10, 0}, {12, 1, 1},
       {36, 0, 0, 1}, {41, 1, 2}, {20, 1, 3},
       {36, 3, 2, 4}, {20, 0, 5}, {33, 5, 4, 6}, {8, 6}},
      Function[{x}, 0]]
```

Here is the graph of g[x].

```
Plot[g[x],{x,-4,4}, PlotRange -> {-5,40} ]
```

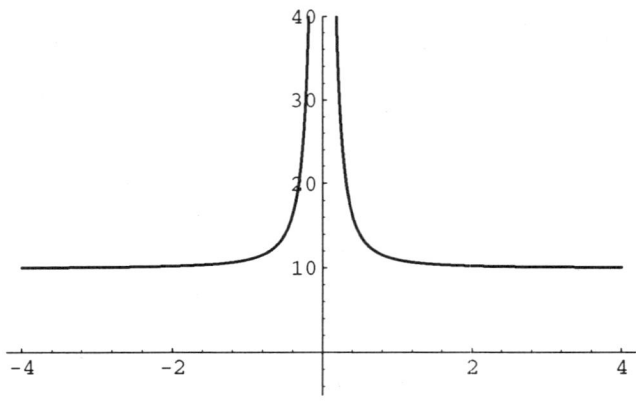

Estimate the values of the horizontal and vertical asymptotes of g, then enter the definition of a function f whose graph will resemble the graph of g. To test your guess, show f and g in the same plot. If you are not satisfied with the result, try again. Change the definition of f as many times as you like to achieve a good fit; a perfect fit is not necessary.

```
Clear[f]
f[x_] := ??
```

6

```
Clear[f]
f[x_] := ??
Plot[ {f[x],g[x]}, {x,-4, 4}, PlotRange -> {-5,40}]
```

4. Here are some more compiled mystery functions. Plot the graph of g (use the trial and error method to find appropriate domain and range for your plot). Then find a rational function f whose graph closely approximates the graph of g, and plot f and g together. Repeat as often as you like. It is possible to produce a function f whose graph is identical with that of g, although this may require considerable time and effort.

a.

Clear[g]

```
g = CompiledFunction[{_Real}, {0, 2, 11, 0},
     {{1, 17}, {4, 1, 0}, {36, 0, 0, 1}, {36, 0, 2},
     {39, 2, 3}, {12, 1, 0}, {20, 0, 4},
     {33, 1, 3, 4, 5}, {36, 0, 0, 6}, {12, -1, 1},
     {20, 1, 7}, {33, 6, 0, 7, 8}, {41, 8, 9},
     {36, 5, 9, 10}, {8, 10}}, Function[{x}, 0]]
```

b.

Clear[g]

```
g = CompiledFunction[{_Real}, {0, 2, 8, 0},
     {{1, 17}, {4, 1, 0}, {12, 1, 0}, {12, -5, 1},
     {20, 1, 1}, {33, 0, 1, 2}, {36, 2, 2, 3}, {41, 3,4},
     {20, 0, 5}, {36, 5, 4, 6}, {33, 0, 6, 7}, {8, 7}},
     Function[{x}, 0]]
```

Mathematica Commands Needed:

```
Apart[ rational expression ]

Clear[ variables ]

Plot[ f[x], {x, a, b}, PlotRange -> { c, d } ]

Plot[{ f[x], g[x] }, {x, a, b}, PlotRange -> { c, d } ]
```

Limits

One way to use the computer when computing limits is to evaluate the limit exactly using techniques like those in sections 2.4 and 2.7 of your textbook, and then to graph the function over an appropriate interval to check the value of the limit you have computed. This will also improve your geometric understanding of limits. If you are unable to compute the value of a limit, the graph can give a reasonable indication of whether the limit exists, and, if it does, you can approximate the value of the limit by evaluating the function.

Exercises

1. Find the value of each of the following limits exactly using the techniques of section 2.4 or 2.7 of your textbook. Show your work in a text cell below each problem. Plot the function over an appropriate interval to check your answer.

    a. $\text{Lim}_{x \to \infty} (9x^2 - 2x + 1)1/2 / x$

    b. $\text{Lim}_{x \to \infty} ((5x^2 - 3x + 1) / (x + 1) - 5x)$

    c. $\text{Lim}_{x \to 0} (1 - \cos(2x)) / x^2$

    d. $\text{Lim}_{x \to \infty} (1 - \cos(2x)) / x^2$

2. None of the following limits can be found using the techniques of sections 2.4 or 2.7. Plot the graph of the function over an appropriate interval to determine if the limit appears to exist. If it does, estimate the value of the limit to three or four decimal places by evaluating the function.

    a. $\text{Lim}_{x \to 0} (3x - 1) / x$

    b. $\text{Lim}_{x \to 0} (1 + 1/x)^x$

    c. $\text{Lim}_{x \to \infty} \text{Log}(x)$

    d. $\text{Lim}_{x \to \infty} \text{Log}(x) / x^{(1/20)}$ Hint: Consider very large values of x before you draw any conclusions.

Mathematica Commands Needed:

```
Clear[ variables ]

Plot[ f[x], {x, a, b} ]
```

Introduction to Derivatives

The following exercises are designed to get you to think about how you might describe the behavior of a given function over an entire interval. You'll aim first for a qualitative description, then a quantitative description. (Use a dictionary to make sure you understand the meaning of each of these terms.) You will also examine how the description changes as the size of the interval varies. The material in section 3.1 of the textbook may help you to do these exercises.

Exercises

1. Below is a plot of the function $f(x) = x^3 - 5x + 1$ for $-2 < x < 2$. Describe the graph. On the whole, would you say this function is increasing or decreasing on the interval $-2 < x < 2$? How could you quantify your answer?

```
Plot[x^3 - 5x + 1 , {x, -2, 2}]
```

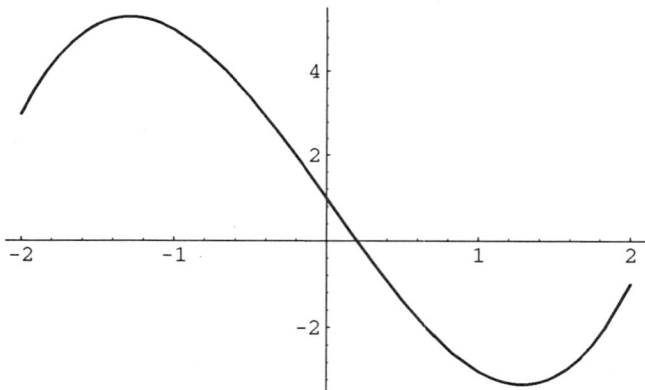

2. Below is a plot of the functions $f(x) = x + \sin(\pi x)$ and $g(x) = (-1/2) x + x^2$ on the interval $-1 < x < 2$. On the whole, both functions seem to be increasing on this interval. Which function increases more, $f(x)$ or $g(x)$? How much more? (The second question asks you to quantify your answer to the first question.)

`Plot[{x + Sin[Pi x], (-1/2) x + x^2}, {x,-1,2}]`

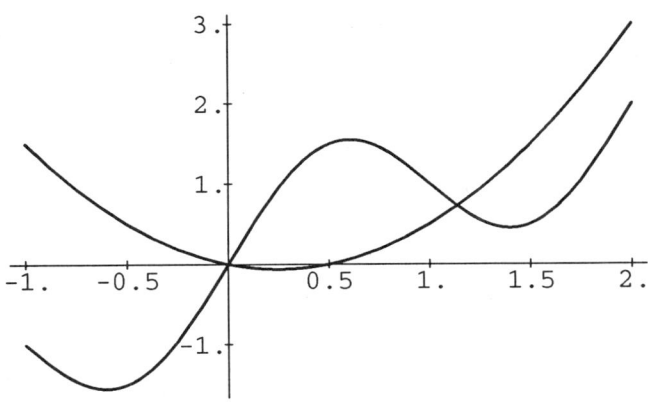

3. a. Plot the graph of the function $f(x) = x^3 - 5x + 1$; first on the interval $-2 \le x \le 2$, then on the interval $-1 \le x \le 1$, and finally for $-0.5 \le x \le 0.5$.

b. Describe the effect on the plot of "zooming in" on x=0.

c. For each of your plots, give the equation of the straight line that you feel best describes the curve over the entire interval. The line you choose should pass through the point (0,1). There is more than one reasonable answer, so justify your particular choice of line. How did you compute the slope of each line?

4. a. Plot the graph of the function $f(x) = \sin(6x)$; first on the interval $-1 \le x \le 1$, then on $-0.5 \le x \le 0.5$, then on $-0.25 \le x \le 0.25$, and finally on $-0.1 \le x \le 0.1$.

b. Describe the effect on the plot of "zooming in" on x=0.

c. For each plot above, give the equation of the straight line that you feel best describes the curve. The line you choose should pass through the point (0,0). There is more than one reasonable answer. Justify your choice. How did you compute the slope of each line?

5. Generalize from your experience in exercises 3 and 4 in order to complete the following loosely stated proposition in your own words.

Proposition: The effect of "zooming in" on the graph of the function f(x) at the point x=0 is to . . .

6. a. Plot the graph of the function $f(x) = (x^2)^{1/3}$; first on the interval $-2 \leq x \leq 2$, then on $-1 \leq x \leq 1$, and finally on $-0.1 \leq x \leq 0.1$.

   b. Describe the effect on the plot of "zooming in" on x=0.

   c. How true is the proposition in exercise 5?

Mathematica Commands Needed:

```
Clear[ variables ]

Plot[ f[x], {x, a, b} ]
```

Tangent Lines

Exercises

1. Plot $f(x) = x^3 - x^2$.

    a.   Find an equation for the tangent line to the curve at a typical point $(a, a^3 - a^2)$. You don't need the computer for this.

    b.   Use the computer to find all points where the tangent line is horizontal.

    c.   Find all points where the tangent line has slope 1.

2. Plot $f(x) = x^3 - x$.

    a.   Find an equation for the tangent line to the curve at a typical point $(a, a^3 - a)$.

    b.   Find all points where the tangent line is horizontal.

    c.   Find all points where the tangent line is parallel to the line $y = 2x + 5$.

    d.   Show the graph of $f(x)$, the line $y = 2x + 5$, and all of the tangent lines in part c on the same plot.

3. Enter the following definitions:

```
f[x_] := x^3
tangentline[f_,a_,x_] := f[a] + f'[a](x-a)
```

    a.   Plot $f(x) = x^3$ and the tangent line to the graph of $f(x)$ at $(a,f(a))$ with different values of the variable a. Note the number of times the tangent line meets the curve.

    b.   Use the **Solve** command to determine if the tangent line always intersects the curve in a point other than (a,a^3).

The remaining exercises use the **tangentline** function defined in exercise 3.

4. a.   Find the x-intercept of the tangent line to the curve $f(x) = x^3 - 2x^2 + x + 1$ through

the point (0,1).

b.   Use the **Solve** command and **tangentline[f,a,x]** to find the x-intercept of the
tangent line to this curve through an arbitrary point $(a, a^3 - 2a^2 + a + 1)$.
You should enter **Clear[a]** first .

c.   Find a formula for the x-intercept of the tangent line to any curve $y = f(x)$ through
an arbitrary point $(a, f(a))$. You do not have to use the computer for this, but if
you do, you should clear both f and a first.

5. a.   Does the tangent line to the curve $y = x^2$ at the point (1,1) pass through the
point (6,12)? Explain.

b.   Find equations of all tangent lines to the curve $y = x^2$ that do pass through the
point (6,12).

6. An astronaut is travelling from left to right along the curve $y = x^2$. When she shuts off
the engine, she will fly off along the tangent line to the curve at the point where she is at
the moment the engine is shut off. At what point should she shut off the engine in order
to reach the point (4,9)?

7. Plot the function $f(x) = x^4 - 8x^2$. Find four points on the curve where the tangent line
to the curve passes through the point (-11/3,49).

BONUS: Experiment with points other than (-11/3,49) to find out how many tangent
lines to the curve $y = x^4 - 8x^2$ pass through any given point. Try points inside the curve
and outside, near the curve and far away. A good intermediate project might be to
classify the points on the x-axis or on the y-axis. Describe the results of your
experiments.

Mathematica Commands Needed:

```
Clear[ variables ]

Plot[{ f[x], g[x] }, {x, a, b}, PlotRange -> { c, d } ]

Solve[ lhs == rhs , variable ]
```

Newton's Method

The following command gives an equation for the tangent line to the curve y = f(x) through the point (a, f(a)). It is not a built-in Mathematica command, so you should enter it each time you open this notebook.

```
tangentline[f_,a_,x_] := f[a] + f'[a](x - a)
```

Here is a plot of the function $g(x) = x^3 - 2x^2 + x + 1$ and the tangent line to this curve through the point (-1,3).

```
g[x_] := x^3 - 2 x^2 + x + 1;
Plot[{g[x],tangentline[g,-1,x]},{x,-2,1}]
```

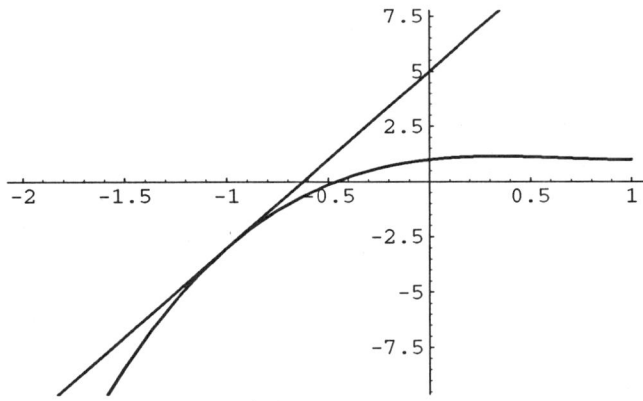

Newton developed a wonderful use of the tangent line. Starting with a value like -1 , which is fairly close to a root of g(x), it is usually the case that the intersection of the tangent line with the x-axis is closer to the root of g(x) than the starting value. In the plot above, notice that the intersection of the tangent line with the x-axis is roughly (-0.6,0), much closer to the actual x-intercept of the graph than is the point (-1,0). You can find the intersection of the tangent line and the x-axis exactly using **Solve**.

```
N[Solve[tangentline[g,-1,x] == 0, x]]

{{x -> -0.625}}
```

The idea of Newton's method for solving the equation g(x) = 0 is to use the x-intercept of the tangent line as an approximate solution of the equation, then to repeat the procedure using a better tangent line. For example, if you find the x-intercept of the tangent line to the graph at (-0.625, g(-0.625)) it should be even closer to the actual solution.

```
Plot[{g[x],tangentline[g,-0.625,x]},{x,-2,1}]
```

```
N[Solve[tangentline[g,-0.625,x]==0,x]]
```

```
{{x -> -0.485786}}
```

Look again at the graph. As you can see, the new x-intercept is pretty close to the actual solution. You can continue applying the procedure to obtain more accurate approximations to the x-intercept of the original curve.

```
N[Solve[tangentline[g,-0.485786,x]==0,x]]
```

```
{{x -> -0.465956}}
```

```
N[Solve[tangentline[g,-0.465956,x]==0,x]]
```

```
{{x -> -0.465571}}
```

```
N[Solve[tangentline[g,-0.465571,x]==0,x]]
```

```
{{x -> -0.465571}}
```

The last two iterations give the same answer, so we'll agree to accept -0.465571 as an approximation to the root that's accurate to six decimal places. Just to make sure, we can evaluate f(-0.465571).

```
f[ -0.465571]
```

$$8.1448 \ 10^{-7}$$

Pretty close to 0!!

Exercises

For each problem below, the first command you enter should be **Clear[f]** to clear the function f from the previous problem.

1. Let $f(x) = x^4 + x - 19$.

a.   Show that f(2) < 0 < f(3). Explain why f(x) must have a root between 2 and 3.

b.   Apply Newton's method, starting with a=2 and iterating the command
     N[Solve[tangentline[f,a,x]==0]], replacing the variable a with the result of the
     previous iteration until the output does not change (accuracy to 5 decimal places).

c.   Evaluate f(a) for your last a. Use **N[f[a],10]**.

2. Remember to **Clear[f]** before you start.

a.   Plot $y = x \; \sin(x)$ for $0 \le x \le \pi$.

b.   Explain why the maximum value of x sin(x) on this interval occurs when
     $\sin(x) + x \cos(x) = 0$.

c.   Solve is only reliable for polynomials of low degree. If you try
     Solve[Sin[x] + x Cos[x] == 0,x] to find the maximum, it won't work. Try and
     see.

d.   Newton's method is a fast, reliable way to approximate solutions to equations
     like $\sin(x) + x \cos(x) = 0$. Use Newton's method, with initial a = $\pi/2$, to
     compute an approximate solution (accurate to five decimal places) to this
     equation.

e.   Estimate the maximum value of x sin(x) on the interval $0 \le x \le \pi$.

3. The equations that must be solved in the problems above are ultimately of the form
b + m(x-a)=0. The solution is x = a - b/m .

a.   Clear f and a and enter **Solve[tangentline[f,a,x]==0]**. Compare the result to the
     formula on page 205 of the textbook called Newton's formula.

b.   Explain the following: **newpoint[a_] := a - f[a]/f'[a]**

c.   Enter the definition of **newpoint**, and repeat problem 2 using **newpoint**. (Try
     newpoint[%] after the first iteration.)

The following exercise illustrates that care must be taken when applying Newton's
method.

4. a.   Apply Newton's method with initial point a = 1/Sqrt[5] to find a solution of the
        equation $x^3 - x = 0$. Do not do more than 10 iterations.
   b.   Explain what is happening in part a using the plot below.

```
Plot[{f[x],tangentline[f,-.447214,x],
tangentline[f,.447214,x]}, {x,-.8,.8}]
```

Read Me !

Mathematica has a program for Newton's method built into its library of standard procedures. The program is executed by entering

```
FindRoot[f[x] == c, {x, x0}, AccuracyGoal -> k]
```

where x0 is the initial approximation to the solution. For example, to find a solution of the equation $\sin(x) + x\cos(x) = 0$ as in exercise 2, we can enter

```
FindRoot[Sin[x] + x Cos[x] == 0, {x,Pi/2},
                                AccuracyGoal -> 6]
```

```
{x -> 2.02876}
```

Challenge:

Write a short program that will compute the solution to an equation using Newton's method. Your program should quit when successive approximations to the solution agree to five digits to the right of the decimal place. You may need to ask for help after you have decided on a general strategy.

Mathematica Commands Needed:

```
Clear[ variables ]

N[ expression, number of decimal places ]

Plot[ f[x] , {x, a, b} ]

Plot[{ f[x], g[x] }, {x, a, b} ]

Solve[ f[x] == a , x ]
```

Derivatives and Estimates I

In this notebook you will investigate how the derivative of a function on an interval governs the behavior of the function itself on the interval. This is an application of the Mean Value Theorem and its corollaries which are discussed in Section 4.1.

Example

Plot the three functions $f(x) = 1/2 \sin(2\pi x)$, $g(x) = x^{1/2} - x^{-1/2}$, and $h(x) = (x - 1)(x - 1/2)$ on the same set of axes for $1 < x < 1.1$.

```
f[x_] := 1/2 Sin[2 Pi x]
g[x_] := Sqrt[x] - 1/Sqrt[x]
h[x_] := (x - 1) (x - 1/2)

Plot[{f[x],g[x],h[x]},{x,1,1.1},
PlotStyle -> {Dashing[{}],Dashing[{.03,.03}],
Dashing[{.01,.03,.03,.03}]}]
```

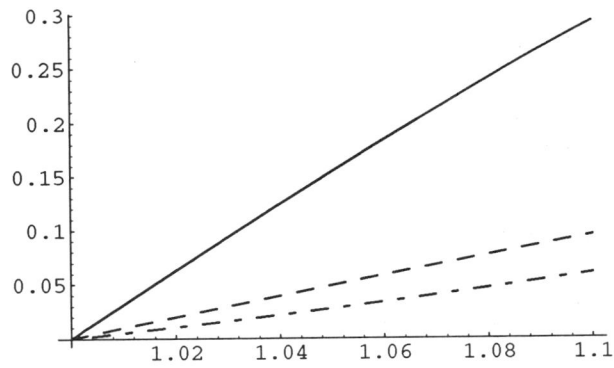

The graph of $f(x)$ is solid, the graph of $g(x)$ is dashed, and the graph of $h(x)$ is dashed with different sized dashes.

The value of the derivative of each of these functions at $x = 1$ can be used to explain the appearance of the plot.

```
f'[1]
```

Pi

g'[1]

1

h'[1]

1
–
2

The fact that the graph of g(x) lies between the graphs of f(x) and h(x) near 1 can be inferred from the fact that f, g, and h all have the same value at 1, and the fact that, since f'(1) > g'(1) > h'(1), the function f is increasing faster than g, and g is increasing faster than h near x=1.

Exercises

1. Plot f(x) = x²/2, g(x) = sin(x) - x cos(x) , and h(x) = -x²/2 on the same axes for 0 < x < 8. Remember to **Clear[f,g, h]**.

As in the example, f is on top, g in the middle, and h on the bottom.

In the example, knowledge of the slope of each graph at the left endpoint was sufficient to explain the appearance of the plot. Here, the situation is just a bit more complicated.

Evaluate the derivatives of each function at x = 0. Since all the derivatives have the same value at x = 0, this value alone does not explain the appearance of the plot.

Plot the derivatives f'(x), g'(x), h'(x) over the entire interval 0 ≤ x ≤ 8, and use this plot of the derivatives to explain the appearance of the original plot of f(x), g(x), and h(x).

2. a.   Complete the statement of the following proposition.

If f(a) = g(a) and f'(x) = g'(x) for all x in the interval from a to b, then the relationship between f(b) and g(b) is ...

b.   Apply a corollary of the Mean Value Theorem to the function f(x) - g(x) to prove that your proposition is true.

3. For each function f(x) given below, plot f'(x) for a ≤ x ≤ b. Use the plot to read off the greatest value, M, of f'(x) for a ≤ x ≤ b and the least value, m, of f'(x) for a ≤ x ≤ b. Use these numbers to define two new functions:

20

```
Pi
```

g'[1]

1

h'[1]

$$\frac{1}{2}$$

The fact that the graph of g(x) lies between the graphs of f(x) and h(x) near 1 can be inferred from the fact that f, g, and h all have the same value at 1, and the fact that, since f'(1) > g'(1) > h'(1), the function f is increasing faster than g, and g is increasing faster than h near x=1.

Exercises

1. Plot $f(x) = x^2/2$, $g(x) = \sin(x) - x\cos(x)$, and $h(x) = -x^2/2$ on the same axes for $0 < x < 8$. Remember to **Clear[f,g, h]**.

As in the example, f is on top, g in the middle, and h on the bottom.

In the example, knowledge of the slope of each graph at the left endpoint was sufficient to explain the appearance of the plot. Here, the situation is just a bit more complicated.

Evaluate the derivatives of each function at x = 0. Since all the derivatives have the same value at x = 0, this value alone does not explain the appearance of the plot.

Plot the derivatives f'(x), g'(x), h'(x) over the entire interval $0 \le x \le 8$, and use this plot of the derivatives to explain the appearance of the original plot of f(x), g(x), and h(x).

2. a.   Complete the statement of the following proposition.

If f(a) = g(a) and f'(x) = g'(x) for all x in the interval from a to b, then the relationship between f(b) and g(b) is ...

    b.   Apply a corollary of the Mean Value Theorem to the function f(x) - g(x) to prove that your proposition is true.

3. For each function f(x) given below, plot f'(x) for a $\le$ x $\le$ b. Use the plot to read off the greatest value, M, of f'(x) for a $\le$ x $\le$ b and the least value, m, of f'(x) for a $\le$ x $\le$ b. Use these numbers to define two new functions:

Derivatives and Estimates II

In the previous exercises, you used the first derivative to trap the graph of a given function between two linear functions. In the following exercises, you'll use the second and third derivatives to approximate a given function by a quadratic or cubic polynomial.

Exercises

1. Since $\text{Limit}_{x \to 0} \sin(x)/x = 1$, we know that $\sin(x)$ and x are very close for x near 0.

   a. Plot the function $\text{error}(x) = \sin(x) - x$ over the interval $-1 \le x \le 1$.

   b. The plot in part a should resemble a cubic curve. Try to fit a curve of the form $y = kx^3$ to the plot of error(x), choosing the constant k to make the space between the two curves as small as possible. Experiment with the **Plot** command below to obtain the "best" value of k that you can to two decimal places. (The error curve will be solid and the test curve will be dashed.)

   ```
   Plot[{error[x], k x^3},{x,-1,1},PlotRange->All,
           PlotStyle->{Dashing[{}],Dashing[{.03,.03}]}]
   ```

   c. Since $\sin(x) = x + \text{error}(x)$, you have just found a k so that $\sin(x)$ is very closely approximated by the third degree polynomial $P(x) = x + k x^3$. In a single plot show $\sin(x)$ and $P(x)$ over the interval $-2\pi \le x \le 2\pi$ using your "best" value of k, and comment on the quality of the approximation you have achieved.

   d. Using your "best" value for k, evaluate the first, second, and third derivatives of both $\sin(x)$ and $P(x) = x + k x^3$ at $x = 0$. Comment on the significance of these calculations.

   e. Compare **N[Sin[1]]** with **N[P[1]]**, using your "best" value for k.

2. Let n be a positive integer. It is a fact that when x is small, $1 + n x$ is very close to $(1 + x)^n$.

   a. Plot $\text{error}(x) = (1 + x)^n - (1+n x)$ with $n = 6$ over the interval $-.01 \le x \le .01$.

   b. The plot in part a should resemble a parabola. Try to fit a curve of the form $y = k x^2$ to the plot of error(x), choosing the constant k to make the space between the two curves as small as possible. Experiment as in exercise 1 to obtain the "best" value of k that you can.

c.   Since $(1+x)^6 = 1 + 6x + error(x)$, you have just found a k so that $(1+x)^6$ is very closely approximated by the second degree polynomial $P(x) = 1 + 6x + k x^2$. Plot $(1+x)^6$ together with $P(x)$ using your "best" value of k and comment on the quality of the approximation you have achieved.

d.   Using your "best" value for k, evaluate the first and second derivatives of both $(1+x)^6$ and $P(x) = 1 + 6 x + k x^2$ at $x = 0$.

e.   Determine how the value for k depends on the 6 in $(1+x)^6$. What value of k will make the polynomial $P(x) = 1 + 7x + k x^2$ as close as possible to $(1+x)^7$? Now, find a formula for k that works for any n.

FACT: A function that has 1st, 2nd, 3rd, ... derivatives may be approximated to any desired accuracy by polynomials!

Mathematica Commands Needed:

```
Clear[ variables ]

Plot[{f[x],g[x],h[x]}, {x, a, b}, PlotStyle -> { Dashing[{}],
                                    Dashing[{.03, .03}]} ]
```

Graphing

For each function below, do the following:

   a.   Plot the function on an interval that seems appropriate.

   b.   Use Mathematica to find all critical points and inflection points. (You may need to revise the range for the plot after you find the critical points.)

   c.   Print the plot only.

   d.   On the printed copy, label all relative maxima, relative minima, global maxima, global minima, horizontal and vertical asymptotes, and inflection points accurate to 5 decimal places.

(Be sure to **Clear[f]** between problems.)

1. $f(x) = (x^2-1)/(x^3-2)$

2. $f(x) = (6 + x - x^2 - x^4)^{1/2}$

3. $f(x) = 4\sin(x) + 2\cos(3x)$

4. $f(x) = 1/(x + \sin(x))$

Mathematica Commands Needed:

```
Clear[ variables ]

Plot[ f[x], {x, a, b}, PlotRange -> { c, d } ]

Solve[ f[x] == a , x ]

f'[x]        ( Compute the derivative. )

f''[x]       ( Compute the second derivative. )
```

Max-Min Problems

The computer can help with max-min word problems in two ways. It can find all of the critical points by solving the equation f'(x) = 0, and it can plot the function over whatever interval is appropriate to the problem so that you can easily see which of the critical points produces the maximum or minimum value. For most of the problems in the book the computer can be very helpful. These additional exercises take advantage of the abilities of the computer to present problems that would be very difficult to do without a computer because of the messy algebra or complex geometry.

Exercises

1. Enter the cell below and you will see a picture of a rectangle inscribed under one hump of the sine function. The value of c determines the location along the x-axis of lower left corner of the rectangle. Try a few values for c.

```
c=.5 ;
Plot[Sin[x], {x,0,Pi}, Prolog ->
{Line[{{c,0},{c,Sin[c]},{Pi-c,Sin[Pi-c]},{Pi-c,0}}]}]
```

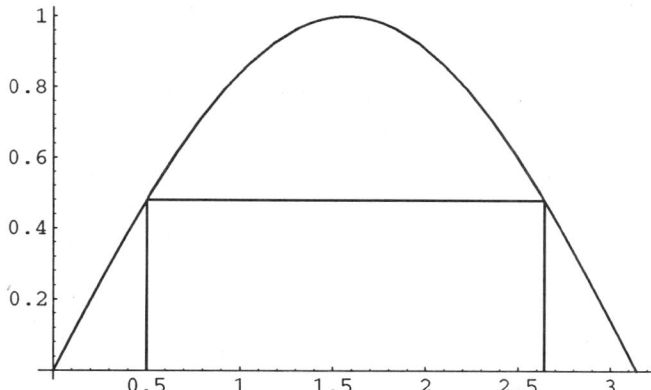

a.   What values of c make sense?

b.   Clear the variable c and define a function A[c] which gives the area of the rectangle drawn above.

c.   Plot the function A[c] over the domain given in part a, and use **FindRoot** on A'[c] to locate the value of c which produces the largest area.

25

2. Plot the function $f(x) = (x+100)(\sin(x/50) + \cos(x/25))/(2x)$ over the interval $100 \le x \le 1000$.

The limit of $f(x)$ as $x \to +\infty$ is difficult to determine. On the other hand, if we eliminate the sines and cosines we are left with $(x+100) / (2x)$, whose limit as $x \to +\infty$ is clearly $1/2$.

    a.   Explain why the values of $\sin(x/50) + \cos(x/25)$ repeat with period 100π.

    b.   Find the maximum value M and the minimum value m of the function $\sin(x/50) + \cos(x/25)$ on the interval $0 \le x \le 100\pi$.

    c.   Plot the three functions $m\ (x+100)/(2\ x)$, $(x+100)\ (\sin(x/50)+\cos(x/25)) / (2\ x)$, and $M\ (x+100)/(2\ x)$ on the interval $100 \le x \le 1000$.

    d.   What can you say about $\lim_{x \to +\infty} (x+100)\ (\sin(x/50) + \cos(x/25))/(2\ x)$? Explain.

3. This problem concerns the use of "solar" energy for long-distance space travel.

Suppose you were to send a spacecraft from Earth to Sirius, our brightest stellar neighbor, emitting over twenty times as much energy as the sun. Suppose also that you wished to build a solar collector large enough to power an on board computer workstation capable of running Mathematica. How large would your solar collector have to be?

Use the following data to answer this question:

    Sirius is located at a distance of 8.7 light-years from the sun (8.2×10^{16} meters).

    The intensity of the energy received from the sun at distance x from the sun is $3.9 * 10^{23} / x^2$ in kilowatts per square meter.

    The intensity of the energy received from Sirius at distance x from Sirius is $9.8 * 10^{24} / x^2$ in kilowatts per square meter.

Assume that your computer workstation uses 1.5 kw (150 watts), and that your solar collector is 100% efficient in converting solar power to usable electric power.

4. The regression line is a common tool used to analyze data. In the simplest case the data consists of points in the x-y plane. The regression line is the line for which the sum of the squares of the vertical distances from the line to the collection of points is

26

minimized. The purpose of this exercise is to see how calculus can be used to get a formula for producing the regression line.

In parts a and b of this problem you are asked to experimentally fit a line to some data points, then to write a report of your activity. Begin by entering the cells below.

Clear[points]

points = { {1,1}, {2,0}, {1.5,3},{4,5}}

a. Look at the picture below, alter the values of the slope m and y-intercept b in the command **regressplot[m , b, points],** and press enter. Mathematica will produce a plot containing the line with the slope m and y-intercept b and it will compute the sum of the squares of the vertical distances from the points to that line. Try to reduce the sum of squares with successive guesses of m and b. Repeat until the sum is less than 8.

regressplot[1,2,points]

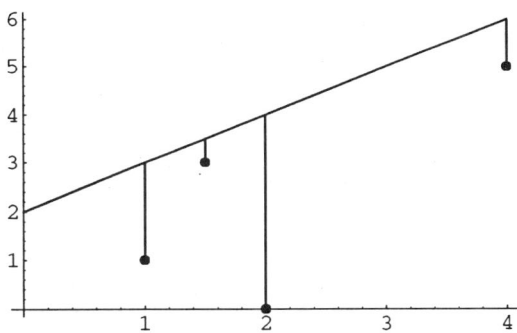

sum of squares = 21.25

b. Repeat the exercise above with this new larger set of points.

Clear[points]

points = { {1.1,3}, {1.5,2}, {1.7,1}, {2,1},{2.3,.89},
{2.5,1}, {2.7,.5}}

When you have achieved a satisfactory fit, write a brief paragraph describing the process by which you arrived at your best line. Where does the problem become difficult -- for example, is a value of 3 for sum of squares in part b easy to achieve? What strategies do you think work best for producing good-fitting lines?

regressplot[-1,2,points]

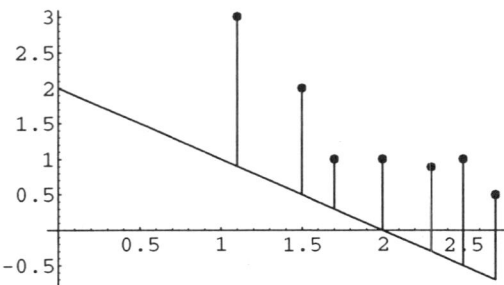

sum of squares = 13.2561

It is a fact that if xav is the average of the x coordinates of the points, and yav is the average of the y coordinates of the points, then the best line passes through the point (xav,yav) and, therefore, we can solve for b in terms of m, making the sum of the squares of the residuals a function with only one variable. We won't prove this fact here, but it does have a natural geometric appeal, even if it is not obvious.

Below you are provided with a Mathematica definition which computes the average of the x coordinates, xav, and the average of the y coordinates, yav, and sets b equal to yav - m xav. The function **squaresum** can then compute the sum of the squares knowing only m and points. Enter this definition.

```
squaresum[m_,points_] := Block[{n=Length[points],
      xav = Sum[points[[i,1]],{i,1,n}]/n,
      yav = Sum[points[[i,2]],{i,1,n}]/n,
      b = yav - m xav },
      {Plus @@ Apply(m #1 + b - #2)^2 &, points, 2] }]
```

Now you can plot the sum of the squares as a function of m.

28

`Plot[squaresum[m,points],{m,-3,3}]`

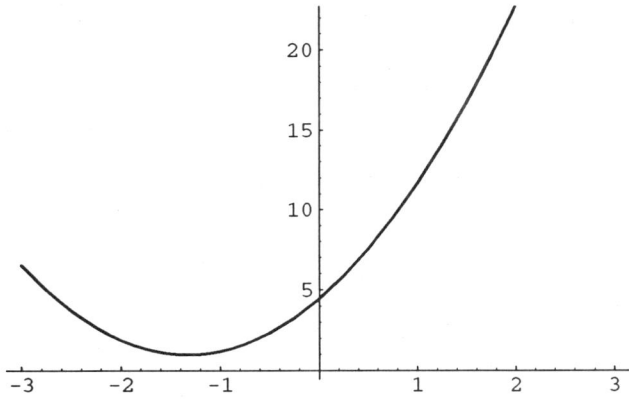

c. The graph above looks like a parabola. Is it?

To find the value of m which produces the minimum evident in the plot above, just use the usual calculus technique of setting the derivative of **squaresum** equal to zero, and solving for the critical point. Since **squaresum** has two variables, m and points, you will need to use the **D[*function, variable*]** command in order to take the derivative with m the variable and points treated as a constant.

`D[squaresum[m,points],m]`

`Solve[% == 0, m]`

How does this result compare to the m you found above through trial and error?

d. Considering only the slope reduces the problem of finding the best-fitting line to a problem involving only one variable, m. The advantage here is that you can get Mathematica to solve the problem generally, obtaining a formula for the optimal slope in terms of the given data.

Replace the specified points listed above with two unspecified points, then solve for the critical slope m.

`Clear[points]`
`points = {{x1,y1}, {x2,y2}}`

e. Repeat the part d, adding a third unspecified point {x3, y3}.

For the curious, here is the definition of the function **regressplot.**

```
regressplot[m_, b_, points_List] :=
  Block[{lastx = Sort[points][[-1,1]]},
    {Show[Graphics[{{PointSize[0.02], Point /@ points},
        {Line /@
           Apply[{{#1, #2}, {#1, m*#1 + b}} & , points, 2]},
         Line[{{0, b}, {lastx, m*lastx + b}}]},
      AxesOrigin -> {0, 0}, Axes -> True]],
    Print["sum of squares = ",
      Apply[Plus, Apply[(#2 - m*#1 - b)^2 & , points, 2]]]}]
```

Mathematica Commands Needed:

```
Clear[ variables ]

D[ expression , variable ]

FindRoot[ f[x] == 0 , { x , a } ]

Plot[ f[x], {x, a, b} ]

Solve[ lhs == rhs , variable ]
```

Some Sums

The curve below is the graph of $f(x) = 1/(1 + x^2)$ over the interval $0 \leq x \leq 2$. The stairstep figure surrounding the curve consists of five rectangles. The area of the stairstep figure is easy to compute as the sum of the areas of the five rectangles and serves to approximate the area under the curve. Together, the picture and caption indicate that the area under the curve is less than 1.27.

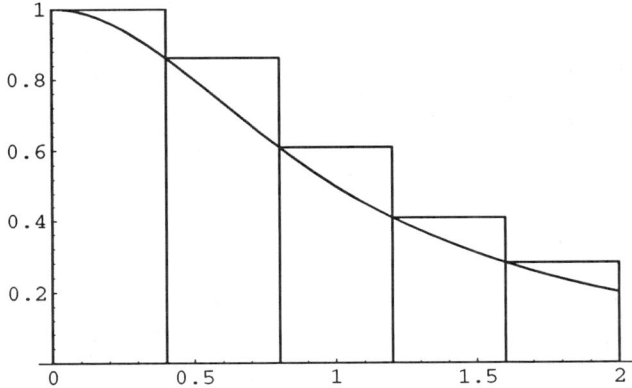

Area of Rectangles = 1.265024002

In this notebook you will learn to use Mathematica to compute sums and to produce pictures containing both curves and rectangles. Focus your attention on how an index variable, typically designated by the letter i, is used to accomplish these tasks.

The Sum Command

Getting Mathematica to sum a sequence of n numbers a[1], a[2], a[3], . . . a[n] is easy. Using the variable i as an index to the terms of the sum, just type the command `Sum[a[i], {i,1,n}]`. Naturally, if you don't give Mathematica more information, it can't do much with the sum. You'll need to specify both the number of terms n and a formula for a[i]. Enter these commands to familiarize yourself with `Sum[ ]`.

```
Clear[a,n];
Sum[a[i], {i,1,n}]

Clear[a,n];
a[i_] := i^2
Sum[a[i], {i,1,n}]
```

31

```
Clear[a,n];
n = 5;
Sum[a[i], {i,1,n}]
```

Exercises

1. Rewrite each of the following expressions concisely using **Sum[ ]**.

a. $1 + 2 + 2^2 + 2^3 + \ldots + 2^{100}$

b. $x^3 + x^4 + x^5 + x^6 + x^7$

c. $1/1^2 + 1/3^2 + 1/5^2 + \ldots + 1/99^2$

d.   x[0]^2 * (x[1] - x[0]) + x[1]^2 * (x[2] - x[1]) + x[2]^2 * (x[3] - x[2])

e.   x[1]^2 * (x[1] - x[0]) + x[2]^2 * (x[2] - x[1]) + x[3]^2 * (x[3] - x[2])

f.   8 x[1]^2 * (x[1] - x[0]) + 8 x[2]^2 * (x[2] - x[1]) + ...
      + 8 x[n]^2 * (x[n] - x[n-1])

Graphic Representation of Sums

Sometimes it's nice to associate a picture to a given sum. You can do this by creating a table of data that Mathematica can use to draw a sequence of rectangles. The variable i will serve as an index to the items in the table, each of which represents a distinct rectangle.

Here's a way to draw a single rectangle having height 3 and base 2 . The function **box[]** takes two points as input and returns a rectangle enclosed by four line segments. The points (a,b) and (c,d) in the function **box[ ]** lie at opposite corners of the rectangle. Thus **box[{0,0},{2,3}]** has opposite corners at (0,0) and (2,3). Try it.

```
box[{a_,b_},{c_,d_}] :=
       Line[{{a,b},{a,d},{c,d},{c,b},{a,b}}]

Show[ Graphics[ box[{0,0},{2,3}] ],
       AspectRatio -> Automatic, Axes -> Automatic ]
```

A table of five rectangles is produced, then shown below. The ith rectangle has height i^2 and base 1.

```
boxes = Table[ box[{i-1,0},{i,i^2}] , {i,1,5}];

Show[ Graphics[ boxes ], Axes -> Automatic ]
```

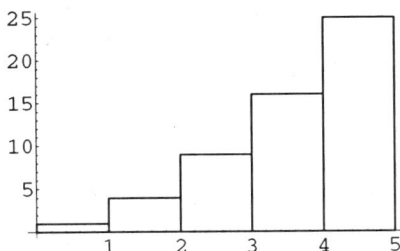

The total area enclosed by these five rectangles is

```
Area = Sum[height[i]*base[i], {i,1,5}]

    = Sum[(i^2)*(1), {i,1,5}]
```

so the picture provides a graphic representation of this sum.

More Exercises

2. a. Explain why the picture produced by the commands below is identical to the picture in the previous example.

```
boxes = Table[ box[{i,0},{i+1,(i+1)^2}] , {i,0,4}];

Show[ Graphics[ boxes ], Axes -> Automatic ]
```

   b. Adjust the table from part a so that the lower left corner of the first rectangle is located at the point (1,0) and the upper right corner of the fifth rectangle is located at (6,25). You may not see how to do everything right away. Try a partial solution and learn from the resulting picture.

3. a. Draw a picture which represents the sum `Sum[(1/2)^i,{i, 1, 4}]`. As before, use rectangles having base 1 and area $(1/2)^i$. Locate the lower left corner of the first rectangle at the point (0,0).

   b. Draw a picture which represents the same sum `Sum[(1/2)^i,{i, 1, 4} ]`. This time, use rectangles which have height 1 and area $(1/2)^i$. The picture will look best if you use rectangles whose lower right corners are at 1, 1/2, 1/4, etc.

c. Evaluate `Sum[(1/2)^i, {i, 1, 8}]` to ten decimal places using `N[ ]`.

d. Adapt your answer to part b to show the sum in part c, then comment on the limit of `Sum[(1/2)^i, {i, 1, n}]` as n approaches infinity.

4. Reproduce the picture at the beginning of this notebook. Recall that $f(x) = 1/(1 + x^2)$ on the interval $0 \le x \le 2$.

HINT: The four corners of the first rectangle are located at (0,0), (0,1), (0.4,0), and (0.4,1) You might write 0.4 as (2/5)*i with i=1. You'll need to use the function f to determine the heights of the remaining rectangles.

```
f[x_] := 1/(1+x^2);

boxes = Table[ box[{??,??},{??,??}], {i,?,?} ];

Plot[f[x],{x,0,2},
        PlotRange ->{0,1}, AxesOrigin ->{0,0},
        Prolog -> {boxes}]

areasum = Sum[f[??]*(?? - ??), {i,?,?}];

Print["Area of Rectangles = ", N[areasum,10]]
```

Mathematica Commands Needed:

`Clear[ variables ]`

`Line[ list of points ]`

`Show[ graphics ]`

`Sum[ function of i , {i, a, b} ]`

`Table[ function of i , {i, a, b} ]`

Riemann Sums

Numerical integration is the process by which estimates of the value of a definite integral are obtained using approximating sums. You have already learned to use Mathematica to evaluate such sums and to draw pictures that can guide your thinking. In this notebook you'll be concerned with determining which type of sum achieves greatest accuracy for a given integral and with measuring the accuracy that is achieved.

First, a quick review of summation notation and the drawing of rectangles.

The function **box[ ]** produces an unfilled rectangle for which two opposite corners have been specified.

```
box[{a_,b_},{c_,d_}] :=
        Line[{{a,b},{a,d},{c,d},{c,b},{a,b}}]
```

Here is a picture of a rectangle with one side on the x-axis, and with its upper left corner touching a curve which lies above the x-axis. Study this example carefully .

```
f[x_] := x^2 + 1
Plot[f[x],{x,0,2},PlotRange ->{0,5}, Prolog ->
                  {box[{.5,f[.5]},{1.5,0}]}]
```

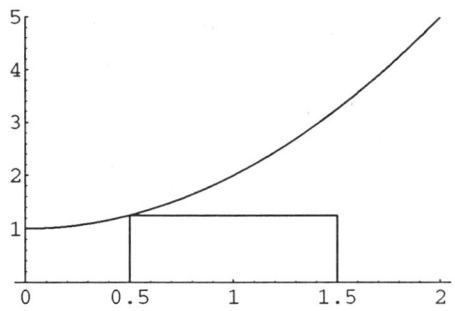

The next goal is to draw a picture of the approximating sum $\Sigma f(c_i)$ (b-a)/n and to compute its value. Use the cell below to define the function, to specify [a,b], to specify the number of intervals in the partition, and finally to define the partition points p[i]. It seems efficient to place all of these commands together in the same cell.

```
Clear[f,n,a,b,p];
f[x_] := Sin[x];
a= 0;
b = Pi;
n = 5;
p[i_] := a + i(b-a)/n;
```

The commands below will enable you to draw a picture of the sum and to compute its value. Enter the cell.

```
leftsumplot := Plot[{f[x],0},{x,a,b},Prolog ->
Table[box[{p[i-1],f[p[i-1]]},{p[i],0}],{i,1,n}]];

leftsum := N[Sum[f[p[i-1]]*(p[i]-p[i-1]),{i,1,n}]];
```

The **Table[]** command from the previous notebook has been incorporated into **leftsumplot**. Note that in **leftsum**, $c_i = p[i-1]$ is the left endpoint of the ith interval of the partition of [a,b].

The setup for this notebook is now complete. To obtain a picture of the left endpoint sum for f(x) on [a,b], just type **leftsumplot** into an input cell and press enter. Type **leftsum** into an input cell and press enter to compute the left endpoint sum itself.

leftsumplot

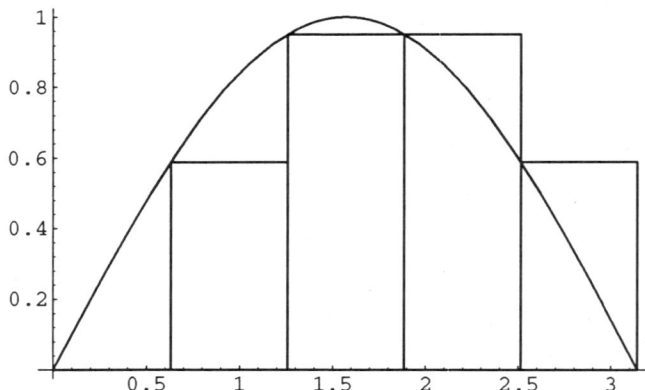

leftsum
1.93377
```
```

Exercises

1. Paste below a copy of the cells which contain the definitions of **leftsumplot** and **leftsum** and make the minor changes necessary to produce plot and sum commands which evaluate the function at right endpoints. Change the name of these new operations to **rightsumplot** and **rightsum**.

2. Repeat exercise 1, evaluating the function at the midpoint of each interval. Call the new operations **midsumplot** and **midsum**.

3. You can get a very accurate numerical estimate of the area under a curve with the command **NIntegrate[f[x],{x,a,b}]**. For each function below, find the left endpoint sum, right endpoint sum, and midpoint sum, including the pictures. Use **NIntegrate[f[x],{x,a,b}]** to examine the accuracy of each estimate. For each problem, be sure to set up the function and partition as in the example. (It would be efficient to paste a copy of the cell from the example, then edit it as necessary.)

   a. $f(x) = x^3 + x - 1$ on the interval $[1,3]$ with $n = 10$.

   b. $f(x) = \sin(x^2 + 1)$ on the interval $[0, 1]$ with $n = 25$.

4. For each statement below, give an example of a function that satisfies the hypothesis, then use your function to draw a picture that illustrates the conclusion of the statement. On the basis of your picture, explain why the statement is true for any function satisfying the given hypothesis.

   a.   If $f'(x) > 0$ for all x in [a,b], then **leftsum** is less than the area under the graph of $f(x)$.

   b.   If $f'(x) > 0$ and $f''(x) = 0$ for all x in [a,b], then **midsum** is equal to the area under the graph of $f(x)$.

   c.   If $f'(x) > 0$ and $f''(x) > 0$ for all x in [a,b], then **leftsum** is more accurate than **rightsum** for computing the area under the graph of $f(x)$.

   d.   BONUS: With the same hypotheses as in part c, determine whether **midsum** is greater than or less than the area under the graph of $f(x)$.

5. You can improve on the qualitative observations in exercise 4 by computing the largest possible error resulting from the use of a particular type of approximating sum. The function $f(x) = 4x^3 - 6x^2 + 7x + 1$ is a positive, increasing function for $0 \leq x \leq 1$.

The reasoning from exercise 4b may be used to show that

> `leftsum` < area under the curve < `rightsum`

a. Draw a `rightsumplot` and a `leftsumplot` for this function using n=5 intervals in your partition.

b. Imagine subtracting `leftsumplot` from `rightsumplot`. Do you see that only the last `rightsum` rectangle and first `leftsum` rectangle remain? Explain why the following formula is valid for any n and any partition p of [0,1].

> $(\text{rightsum} - \text{leftsum}) = f(1) (p[n]-p[n-1]) - f(0) (p[1]-p[0])$

If the partition intervals have equal length, this result may be written simply as

> $(\text{rightsum} - \text{leftsum}) = (f(1) - f(0)) / n.$

c. It is a fact that the midpoint sum for a positive, increasing function satisfies

> $|\text{midsum} - \text{area under the curve}| \leq (\text{rightsum} - \text{leftsum}).$

Suppose you need to guarantee that for $0 \leq x \leq 1$ the error in **midsum** for the area under the curve f(x) is at most 0.001. How many intervals should you use for your partition?

d. BONUS: Suppose that f(x) is increasing on the interval [0,1] and decreasing on the interval [1,2] with f(0) = 5, f(1) = 6, and f(2) = 1. If you partition [0,2] into 2n intervals of equal length, how large should n be in order to guarantee that the error in the midpoint estimate is at most 0.001?

Mathematica Commands Needed:

```
Clear[ variables ]

Plot[ f[x], {x, a, b}, PlotRange -> { c, d },
            Prolog -> { graphics commands } ]

NIntegrate[ f[x], { x, a, b }
```

Tower of Powers

Mathematica is a wonderful tool for exploration, especially the exploration of functions with complex algebra, and interesting graphs. A perfect example is the sequence of functions x^x, x^{x^x}, $x^{x^{x^x}}$, ... The graphs of these functions on the interval $0 \leq x \leq 1$ have some very intriguing properties. Plot several of them and consider the following:

1. There is a difference between $(((x^{\wedge}x)^{\wedge}x)^{\wedge}x)^{\wedge}x$, and $x^{\wedge}(x^{\wedge}(x^{\wedge}(x^{\wedge}x)))$. Which one do you get if you let $f[x\_] := x^{\wedge}x^{\wedge}x^{\wedge}x^{\wedge}x$?

2. There is a difference between the graphs depending on whether the number of x's is even or odd. None of these functions are defined at zero but it is possible to show that for an even number of x's the limit as x goes to 0 is 1, and for an odd number of x's the limit is 0. The graphs certainly indicate this.

3. The function x^x has a critical point. Can you find its coordinates exactly?

4. Here is a Mathematica definition that will allow you to explore the functions with a very large number of x's.
    `tower[x_,n_] := Nest[x^# &, x, n-1]`
    Enter `tower[x, n]` for a few small values of n, but don't get carried away; it can take a long time to print the result for large n. Surprisingly, it does not take the computer very long to plot the function for large n. Plot `tower[x,100]` and `tower[x,101]` together over the interval $0 \leq x \leq 1$.

5. Notice that even for large values of n, when n is even the function appears to have a critical point, and when n is odd it does not. Also, the coordinates of the critical points appear to be converging to some very interesting values involving e. Try to figure out what they are using `FindMinimum[t[x,n],{x,{a,b}}]`. This particular form of `FindMinimum` is suggested because it does not use the derivative of `tower[x,n]`. Compute the derivative for a few small values of n and you will see why it should be avoided.

6. It is surprising that such a clear pattern in the graphs emerges as n increases considering the potential for round-off error when n is in the 100's or even 1000's.

A Minimal List of Mathematica Commands

Functions

```
Abs[], Sqrt[], Exp[], Sin[], Cos[], Sec[], Tan[], Csc[], Cot[], Log[]
```

Algebra

```
Apart[x/(x^2-4)]
Factor[x^2 - 1]
FindRoot[x^3 + x^2 == x - 1 , {x,1}]
Solve[x^3 + x^2 == x - 1, x]
N[2*Pi,100]
```

Math

```
D[f, x]
Integrate[f, x]
Integrate[f, {x, 0, 1}],
NIntegrate[f, {x, 0, 1}]
Limit[Sin[x]/x, x-> 0 ]
Plot[f, {x, -5, 5}]
Plot[f, {x, -5, 5}, PlotRange->{-3,3}]
```

Constants

```
E, I, Pi
```

Arithmetic Operations

+	plus
–	minus
*	times
/	divided by
^	raised to a power
!	factorial

Special

%	previous output
%n	output number n
{}	used for lists only
()	used in algebraic expressions only
[]	used around function arguments
f[x_]:= expr	defines a function f according to the rule in the expression expr
==	used for equations
=	used to set the value of a constant